Machines and Motion

Wedges

by Erika S. Manley

Bullfrog Books

Ideas for Parents and Teachers

Bullfrog Books let children practice reading informational text at the earliest reading levels. Repetition, familiar words, and photo labels support early readers.

Before Reading
- Discuss the cover photo. What does it tell them?
- Look at the picture glossary together. Read and discuss the words.

Read the Book
- "Walk" through the book and look at the photos. Let the child ask questions. Point out the photo labels.
- Read the book to the child, or have him or her read independently.

After Reading
- Prompt the child to think more. Ask: Wedges are everywhere. Where do you see them? How do they help you?

Bullfrog Books are published by Jump!
5357 Penn Avenue South
Minneapolis, MN 55419
www.jumplibrary.com

Copyright © 2019 Jump! International copyright reserved in all countries. No part of this book may be reproduced in any form without written permission from the publisher.

Library of Congress Cataloging-in-Publication Data

Names: Manley, Erika S., author.
Title: Wedges / by Erika S. Manley.
Description: Minneapolis, MN : Jump!, Inc., [2018]
Series: Machines and motion
"Bullfrog Books are published by Jump!"
Audience: Ages 5–8. | Audience: K to grade 3.
Includes bibliographical references and index.
Identifiers: LCCN 2017052362 (print)
LCCN 2017052902 (ebook)
ISBN 9781624968624 (ebook)
ISBN 9781624968600 (hardcover : alk. paper)
ISBN 9781624968617 (pbk.)
Subjects: LCSH: Wedges—Juvenile literature.
Simple machines—Juvenile literature.
Classification: LCC TJ1201.W44 (ebook)
LCC TJ1201.W44 M347 2018 (print) | DDC 621.8/11—dc23
LC record available at https://lccn.loc.gov/2017052362

Editor: Kristine Spanier
Book Designer: Molly Ballanger

Photo Credits: balipadma/Shutterstock, cover; Hans Christiansson/Shutterstock, 1; rodimov/Shutterstock, 3; andresr/iStock, 4; adaptice photography/Shutterstock, 5, 23tr; robcruse/iStock, 6–7, 23br; Jeff Cleveland/Shutterstock, 8, 23mr; invisible163/Shutterstock, 9 (bulletin board); Joanna Dorota/Shutterstock, 9tl; Aleksandar Mijatovic/Shutterstock, 9tr; Ragnarock/Shutterstock, 9b; True Fake/Shutterstock, 10–11; MichaelJayBerlin/Shutterstock, 11; age fotostock/Superstock, 12; CatherineL-Prod/iStock, 12–13; Ivan Smuk/Shutterstock, 14–15, 23bl; Ugorenkov Aleksandr/Shutterstock, 16; Richard Drury/Getty, 17, 23tl; sergeyryzhov/iStock, 18–19, 23ml; KidStock/Getty, 20–21; Diana Taliun/Shutterstock, 24.

Printed in the United States of America at Corporate Graphics in North Mankato, Minnesota.

Table of Contents

Working with Wedges 4
Build a Wedge 22
Picture Glossary 23
Index 24
To Learn More 24

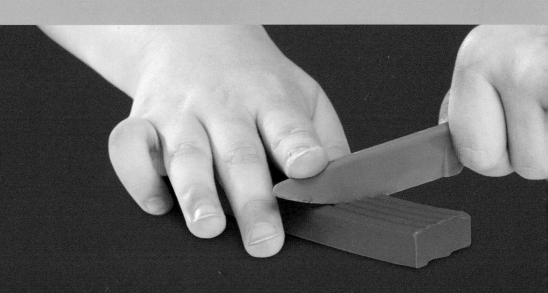

Working with Wedges

What is a wedge?

It is a simple machine.
It holds things in place.

wedge

One side is thin.
It fits in a tight space.
It holds the door.

A tack is a wedge.

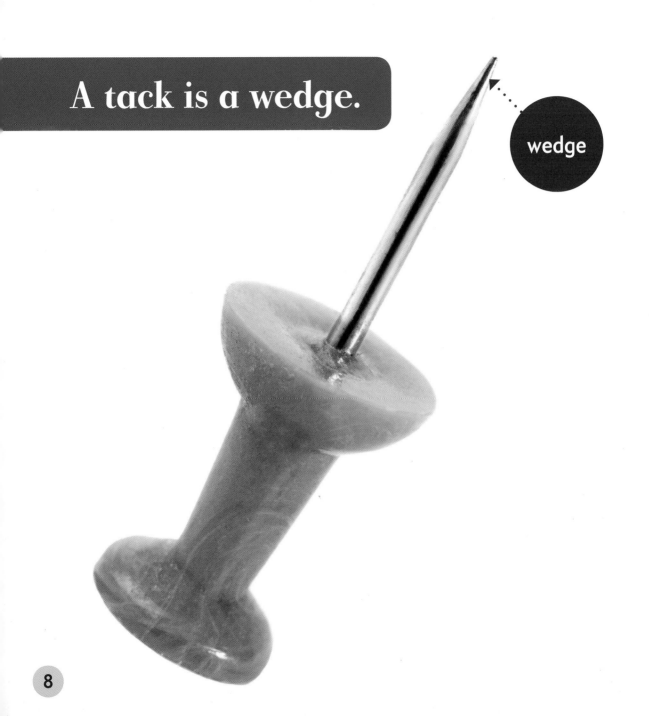

wedge

The art stays up. Pretty!

staple

A staple is a wedge.
Paper stays in place.

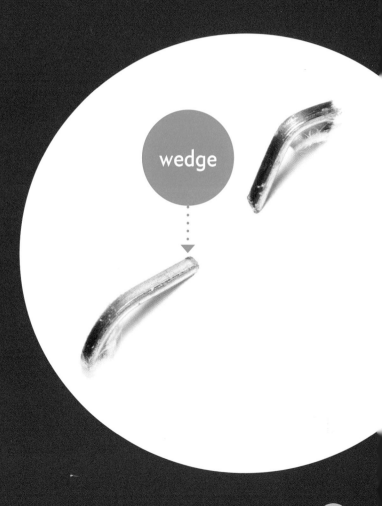

wedge

A wedge does another job.
It can push things apart.
Jay hits the wedge.
The wood splits.
Wow!

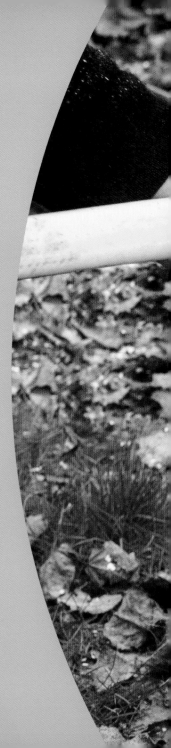

plow

A plow is a wedge.
It moves snow.

Ed uses a wedge.

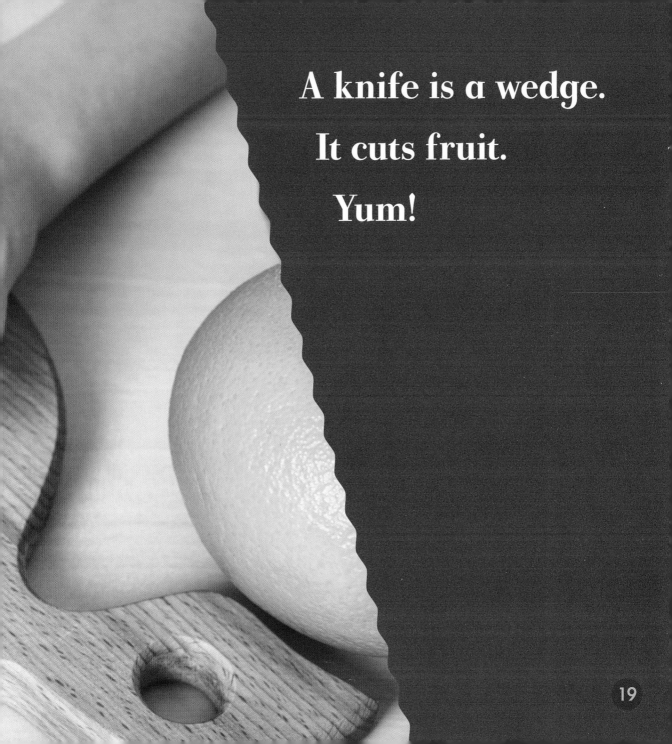

A knife is a wedge.

It cuts fruit.

Yum!

A wedge helps.

Do you use a wedge?

Build a Wedge

Build your own wedge to hold a door open.

You will need:
- scissors
- a cardboard box
- duct tape

Directions:
1. Use the scissors to cut three evenly-sized rectangles from the cardboard box. Ask an adult for help.
2. Stack the cardboard pieces evenly on top of one another.
3. Wrap tape tightly several times at one end of the stacked cardboard pieces. That end of the cardboard should be much thinner than the other side.
4. Place the taped edge under a door. Does the door stay in place? You are using a wedge!

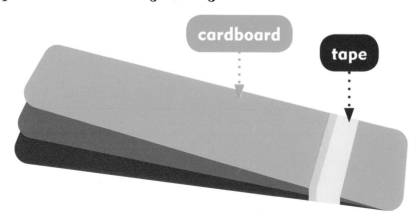

Picture Glossary

carves
Cuts wood, stone, or another hard material into a particular shape.

simple machine
A tool used to make work easier, such as an inclined plane, lever, pulley, screw, wedge, or wheel and axle.

knife
A utensil used to cut food.

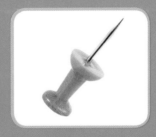
tack
A small nail with a sharp point and a large, flat head.

plow
A blade attached to a vehicle that pushes aside snow or other matter.

wedge
An object that is thin at one end and thick at the other to hold something in place or push things apart.

Index

apart 12
carves 17
cuts 19
holds 5, 6
knife 19
moves 15

simple machine 5
splits 12
staple 11
stays 9, 11
tack 8
thin 6

To Learn More

Learning more is as easy as 1, 2, 3.

1) Go to www.factsurfer.com

2) Enter "wedges" into the search box.

3) Click the "Surf" button to see a list of websites.

With factsurfer.com, finding more information is just a click away.